CROP ROTATION STRATEGIES:
MAXIMIZING YIELD

SMILE WELLBECK

GRATITUDE

I hope this note finds you well and filled with excitement about diving into the world of crop rotation strategies. First and foremost, I want to express my heartfelt gratitude for choosing my book as your guide on this agricultural journey.

I wrote this book with the intention of making crop rotation strategies easy to understand and implement for farmers, gardeners, and anyone interested in sustainable agriculture. Whether you're a seasoned pro or just starting out, I believe there's something valuable for everyone within these pages.

By exploring the principles and practices of crop rotation, you're not only investing in your own knowledge but also in the health of our planet. Crop rotation isn't just about growing better crops; it's about nurturing the soil, minimizing pests and diseases, and promoting biodiversity. It's a holistic approach to farming that benefits both farmers and the environment.

As you delve into the chapters, I encourage you to keep an open mind and embrace the possibilities that crop rotation offers. Experiment, ask questions, and most importantly, enjoy the process of learning and growing alongside your crops.

Once again, thank you for choosing "Crop Rotation Strategies." I'm truly honored to be a part of your agricultural journey.

Copyright © 2024, Smile Wellbeck.

This work and its content are protected under international copyright laws.

No part of this publication may be reproduced, distributed, or transmitted in any form or by any means, including photocopying, recording, or other electronic or mechanical methods, without the prior written permission of the author, except in the case of brief quotations embodied in critical reviews and certain other noncommercial uses permitted by copyright law.

TABLE OF CONTENTS

ABOUT THE BOOK

INTRODUCTION
- UNDERSTANDING CROP ROTATION
- HISTORY AND EVOLUTION OF CROP ROTATION PRACTICES
- IMPORTANCE OF CROP ROTATION IN SUSTAINABLE AGRICULTURE

CHAPTER 1: PRINCIPLES OF CROP ROTATION
1.0 CROP ROTATION BASICS
1.1 CROP DIVERSITY AND ROTATION CYCLES
1.2 SOIL HEALTH AND NUTRIENT MANAGEMENT IN CROP ROTATION
1.3 PEST AND DISEASE MANAGEMENT THROUGH CROP ROTATION

CHAPTER 2: TRADITIONAL AND MODERN CROP ROTATION SYSTEMS
2.0 TRADITIONAL CROP ROTATION PRACTICES
2.1 THREE-FIELD SYSTEM AND OTHER HISTORICAL METHODS

2.2 MODERN ADAPTATIONS AND INNOVATIONS IN CROP ROTATION

CHAPTER 3: PLANNING AND IMPLEMENTING CROP ROTATION
3.0 FACTORS INFLUENCING CROP ROTATION DECISIONS
3.1 CROP SELECTION CRITERIA FOR ROTATION SYSTEMS
3.2 DESIGNING CROP ROTATION SEQUENCES FOR MAXIMUM BENEFIT

CHAPTER 4: CROP ROTATION STRATEGIES FOR DIFFERENT AGROECOSYSTEMS
4.0 CROP ROTATION IN INTENSIVE AGRICULTURE
4.1 CROP ROTATION IN ORGANIC FARMING SYSTEMS
4.2 CROP ROTATION IN PERENNIAL CROPPING SYSTEMS

CHAPTER 5: INTEGRATING LIVESTOCK IN CROP ROTATION
5.0 ROLE OF LIVESTOCK IN SUSTAINABLE CROP ROTATION

5.1 GRAZING AND FORAGE MANAGEMENT IN CROP ROTATION SYSTEMS
5.2 MANURE MANAGEMENT AND NUTRIENT CYCLING

CHAPTER 6: CASE STUDIES AND SUCCESS STORIES
6.0 CASE STUDIES OF EFFECTIVE CROP ROTATION PRACTICES
6.1 SUCCESS STORIES FROM FARMERS IMPLEMENTING CROP ROTATION STRATEGIES
6.2 CHALLENGES AND LESSONS LEARNED

CONCLUSION
THE FUTURE OF CROP ROTATION: INNOVATIONS AND OPPORTUNITIES

APPENDICES
- CROP ROTATION PLANNING WORKSHEETS AND TOOLS

ABOUT THIS GUIDE

Welcome to our guide on crop rotation strategies. We're thrilled to have you on board as we explore this fascinating topic together. Whether you're a seasoned farmer or just starting out with your backyard garden, understanding the ins and outs of crop rotation can make a world of difference in your harvests.

In this guide, we've broken down everything you need to know about crop rotation in simple, easy-to-understand language. No need to worry about complex agricultural jargon here; we're all about keeping things down-to-earth and relatable.

So whether you're a seasoned gardener or just getting started, get ready to take your growing game to the next level.

Let's get planting!

INTRODUCTION

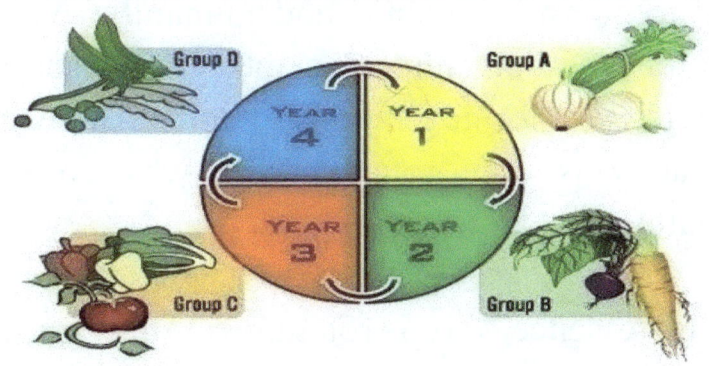

UNDERSTANDING CROP ROTATION

Imagine a lush field, vibrant with the colors of various crops swaying gently in the breeze. Now, picture that same field barren and depleted, its soil exhausted, unable to support life. This stark contrast embodies the essence of crop rotation; a practice as ancient as agriculture itself, yet as vital today as it was centuries ago.

Crop rotation isn't just a farming technique; it's a philosophy deeply rooted in the intricate dance between nature and humanity. At its core, it's about nurturing the land, ensuring its fertility for generations to come. But what exactly is crop rotation, and why does it matter?

In essence, crop rotation is the art of strategically alternating the types of crops grown in a particular field over successive seasons. It's about diversifying, not just for the sake of variety, but to enrich the soil, ward off pests, and promote overall plant health. It's a symphony of sustainability, where each crop plays its part in a harmonious cycle of growth and regeneration.

Think of it as nature's own recipe for soil rejuvenation. Just as we rotate our diets to ensure balanced nutrition, crop rotation replenishes the soil with essential nutrients, preventing depletion and maintaining its vitality. By planting different crops with varying nutrient needs, such as nitrogen-fixing legumes followed by heavy feeders like corn or potatoes, farmers can harness the power of nature to enhance soil fertility without relying on chemical fertilizers.

But crop rotation is more than just a practical solution; it's a testament to the wisdom of generations past. Throughout history, farmers have observed the benefits of rotating crops, often unknowingly tapping into the intricate web of ecological relationships that sustain life on earth. From the ancient civilizations of Mesopotamia to the agrarian communities of medieval Europe, crop rotation has been a cornerstone of agricultural success, passed down through the ages like a treasured heirloom.

Today, in an era marked by environmental uncertainty and agricultural challenges, the relevance of crop rotation has never been greater. As climate change disrupts traditional farming practices and soil degradation threatens food security, crop rotation emerges as a beacon of hope; a time-tested solution to modern-day problems.
Perhaps the most compelling aspect of crop rotation is its inherent resilience. In a world where monoculture dominates vast swathes of farmland, leaving ecosystems vulnerable to collapse, crop rotation offers a lifeline – a path towards regenerative agriculture that honors the interconnectedness of all living things.

So, as we embark on this journey to understand crop rotation, let us not merely see it as a farming technique, but as a profound philosophy that speaks to our relationship with the land and with each other. Let us embrace its principles, not out of necessity, but out of reverence for the wisdom of nature and the legacy of those who came before us.

In the pages that follow, we will delve deeper into the intricacies of crop rotation; exploring its history, its principles, and its practical applications. But before we embark on this exploration, let us pause to reflect on the profound truth at the heart of crop rotation: that by working in harmony with nature, we can cultivate abundance for ourselves and for future generations.

HISTORY AND EVOLUTION OF CROP ROTATION PRACTICES

Close your eyes and journey back in time to the fertile plains of ancient civilizations. Here, amidst the whispers of the wind and the rustle of golden grain, lies the genesis of a practice that would shape the course of agriculture for centuries to come: crop rotation.

Our ancestors, tilling the soil with weathered hands and hearts brimming with reverence for the land, were the pioneers of crop rotation. They observed the natural rhythms of the earth, intuitively understanding the need to replenish the soil and safeguard their harvests against the whims of nature. Thus, the seeds of crop rotation were sown; a testament to human ingenuity and the symbiotic relationship between humanity and the earth.

From the fertile crescent of Mesopotamia to the banks of the Nile River in ancient Egypt, early civilizations practiced rudimentary forms of crop rotation, alternating between crops such as barley, wheat, and legumes to maximize yields and preserve soil fertility. These ancient farmers, guided by trial and error, laid the foundation for a tradition that would endure through the ages.

As civilizations flourished and empires rose and fell, the practice of crop rotation evolved, adapting to the ever-changing needs of agrarian societies. In medieval Europe, where feudal lords held sway over vast estates, crop rotation emerged as a means of both sustenance and social order. The three-field system, a hallmark of medieval agriculture, divided fields into three sections: one for winter grains, one for spring grains, and one left fallow to rest and regenerate; a delicate dance of productivity and renewal.

It was perhaps during the agricultural revolution of the 18th and 19th centuries that crop rotation truly came into its own. Visionaries like Jethro Tull and Charles Townshend championed the scientific principles of crop rotation, recognizing its potential to revolutionize farming practices and alleviate food shortages plaguing burgeoning populations. Their insights paved the way for modern agricultural practices, laying the groundwork for the sustainable farming methods we rely on today.

Despite its centuries-old legacy, the story of crop rotation is far from over. In an era defined by climate change, soil degradation, and food insecurity, the need for sustainable agricultural practices has never been more pressing. Crop rotation, with its proven ability to replenish soil nutrients, suppress pests and diseases, and mitigate the impacts of climate variability, stands as a beacon of hope in a world grappling with environmental challenges.

As we stand at the crossroads of history, let us heed the lessons of the past and embrace the wisdom of crop rotation. Let us honor the resilience of our ancestors, who, with humility and reverence, nurtured the land that sustains us all. And let us forge a new chapter in the epic saga of agriculture; one guided by the timeless principles of sustainability, stewardship, and respect for the earth.

In the chapters that follow, we will delve deeper into the intricacies of crop rotation; exploring its principles, its practical applications, and its transformative potential in an ever-changing world. But for now, let us pause to marvel at the rich tapestry of history that has brought us to this moment; a moment ripe with possibility, where the seeds of change are ready to take root and flourish.

IMPORTANCE OF CROP ROTATION IN SUSTAINABLE AGRICULTURE

Step outside on a crisp morning, and let the sun warm your face as you gaze upon the fields stretching out before you. What do you see? A patchwork quilt of green and gold, teeming with life and possibility. This vibrant tapestry of nature's bounty is a testament to the timeless practice of crop rotation; a cornerstone of sustainable agriculture that holds the key to our future.

At its essence, crop rotation is a dance of diversity – a symphony of plants, soil, and microorganisms working in harmony to nourish the land and sustain life. It's a simple yet profound concept: instead of planting the same crop year after year, farmers rotate between different types of crops, each with its own unique set of needs and contributions to the soil.

Why does crop rotation matter, you may ask? The answer lies in the very fabric of our existence; the soil beneath our feet. Soil is not just dirt; it's a living, breathing ecosystem teeming with millions of microscopic organisms that form the foundation of all life on earth. And just like us, soil needs nourishment to thrive.

Crop rotation is nature's way of replenishing the soil, like a gardener tending to a beloved garden. By rotating crops, farmers can harness the power of nature to naturally replenish soil nutrients, reduce the buildup of pests and diseases, and improve overall soil structure. It's a virtuous cycle of regeneration, where each crop contributes to the health and vitality of the land for generations to come.

The importance of crop rotation extends far beyond the confines of individual fields. In an era defined by climate change, soil degradation, and dwindling natural resources, sustainable agriculture has never been more critical. Crop rotation offers a lifeline – a beacon of hope in a world grappling with environmental challenges.

By diversifying crops and fostering healthy soil ecosystems, crop rotation helps farmers adapt to changing environmental conditions, mitigate the impacts of climate variability, and build resilience against pests, diseases, and drought. It's a practical solution rooted in the wisdom of nature, offering a path towards a more sustainable and resilient food system for all.

Perhaps the most compelling aspect of crop rotation is its profound impact on communities and ecosystems around the world. By promoting biodiversity, preserving soil health, and reducing the need for harmful chemical inputs, crop rotation fosters thriving ecosystems where plants, animals, and humans can coexist in harmony.

As stewards of the land, it is our responsibility to honor and protect the precious resources entrusted to our care. Crop rotation offers a blueprint for sustainable agriculture – a roadmap to a future where abundance, resilience, and harmony reign supreme. Let us embrace its principles with humility and gratitude, knowing that by nurturing the earth, we nurture ourselves and all life on this planet.

In the chapters that follow, we will explore the intricacies of crop rotation; from its historical roots to its practical applications in modern agriculture. But for now, let us pause to marvel at the beauty and resilience of the natural world, and the boundless potential of crop rotation to sustain us all.

CHAPTER 1: PRINCIPLES OF CROP ROTATION

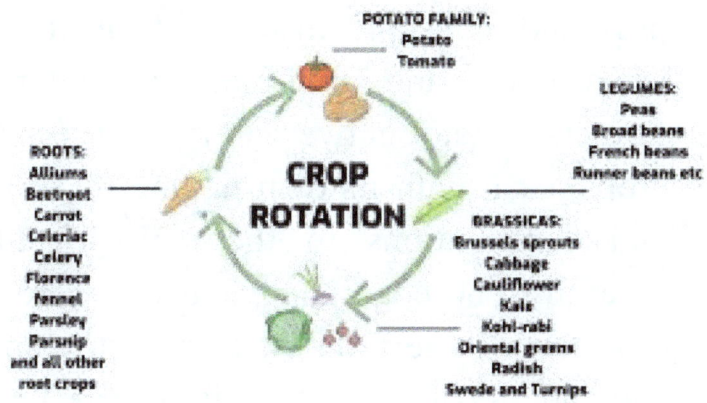

1.0 CROP ROTATION BASICS

Imagine yourself as a gardener, standing before a plot of land bursting with potential. As you survey the earth beneath your feet, you ponder the best way to nurture your plants and ensure a plentiful harvest. This is where crop rotation comes into play; a fundamental principle of sustainable agriculture that harnesses the power of diversity to promote soil health and optimize crop yields.

At its core, crop rotation is a simple yet profound concept: instead of planting the same crop in the same place year after year, farmers rotate between different types of crops, each with its own unique set of needs and contributions to the soil. It's like a carefully choreographed dance, where each crop takes its turn on the stage, enriching the soil and preparing the way for the next act.

Why is crop rotation so important? The answer lies in the delicate balance of nature's ecosystems. Just as our bodies thrive on a diverse diet, so too does the soil depend on a variety of plants to maintain its fertility and resilience. When we plant the same crop repeatedly in the same spot, the soil becomes depleted of essential nutrients, leading to decreased yields and increased susceptibility to pests and diseases.

Crop rotation interrupts this cycle of depletion by replenishing the soil with a rich tapestry of nutrients and organic matter. For example, legumes like peas and beans have the remarkable ability to fix nitrogen from the air and convert it into a form that plants can use. By rotating legumes with nitrogen-hungry crops like corn or lettuce, farmers can naturally boost soil fertility without relying on synthetic fertilizers.

Crop rotation isn't just about soil health; it's also about pest and disease management. Certain crops attract specific pests and diseases, which can build up in the soil over time if not properly managed. By rotating crops, farmers can disrupt pest and disease cycles, reducing the need for chemical pesticides and safeguarding the long-term health of their crops.

Perhaps most importantly, crop rotation is a testament to the interconnectedness of all living things. When we embrace diversity in our fields, we create vibrant ecosystems where plants, animals, and microorganisms can thrive together in harmony. This diversity not only fosters resilience against environmental stresses but also promotes a more balanced and sustainable approach to farming.

So, as you stand before your plot of land, ready to sow the seeds of the future, remember the timeless wisdom of crop rotation. Embrace diversity, nurture the soil, and cultivate a legacy of abundance for generations to come. By practicing crop rotation, you become not just a farmer, but a steward of the land; a guardian of the earth's precious resources and a champion of sustainable agriculture.

1.1 CROP DIVERSITY AND ROTATION CYCLES

Picture a garden bursting with an array of colors, textures, and scents; a symphony of life in full bloom. This vibrant tapestry of diversity is not just a feast for the senses; it's also the secret ingredient behind the success of crop rotation cycles. By embracing diversity in our fields and rotating crops with intention and care, we unlock the full potential of the soil and cultivate a legacy of abundance for generations to come.

At its core, crop diversity is the key to a resilient and sustainable food system. Just as a diverse portfolio protects our investments from market fluctuations, so too does a diverse array of crops safeguard our fields against environmental stresses and uncertainties. When we rotate between different types of crops, each with its own unique set of needs and contributions to the soil, we create a dynamic ecosystem where plants, animals, and microorganisms can thrive together in harmony.

What exactly does crop diversity look like in practice? It's more than just planting a variety of crops; it's about understanding the intricate web of relationships that exist between plants and their environment. It's about selecting crops that complement each other, both in terms of their nutrient needs and their ability to suppress pests and diseases. And it's about embracing the wisdom of nature, which has been cultivating diversity in our fields for millennia.

Crop rotation cycles, then, are the rhythm of this dance of diversity – the heartbeat that sustains life in the soil and ensures a bountiful harvest year after year. By rotating crops in carefully planned sequences, farmers can optimize soil fertility, minimize pest and disease pressure, and maximize overall productivity. It's a delicate balance, to be sure, but one that yields rich rewards for those who are willing to embrace it.

Take, for example, the classic three-field rotation cycle practiced by medieval farmers. In this system, one field would be planted with a nitrogen-fixing legume like clover or alfalfa to replenish the soil with essential nutrients. The second field would be sown with a cereal crop like wheat or barley, while the third field would lie fallow, allowing the soil to rest and regenerate. This simple yet effective rotation cycle ensured a steady supply of food and fodder for both humans and animals, while maintaining the long-term health of the soil.

Today, as we confront the challenges of climate change, soil degradation, and food insecurity, the importance of crop diversity and rotation cycles has never been greater. By embracing diversity in our fields and rotating crops with intention and care, we can build resilience against environmental stresses, mitigate the impacts of climate variability, and foster a more balanced and sustainable approach to farming.

So, as you step out into your fields, ready to sow the seeds of the future, remember the timeless wisdom of crop diversity and rotation cycles. Embrace the richness of nature's tapestry, nurture the soil, and cultivate a legacy of abundance for generations to come.

1.2 SOIL HEALTH AND NUTRIENT MANAGEMENT IN CROP ROTATION

Imagine yourself as a caretaker of the land, entrusted with the stewardship of a precious resource; the soil beneath your feet. As you dig your hands into the earth, you can feel its pulse, its life force flowing through your fingers. This soil, this humble foundation of all life on earth, is more than just dirt; it's a living, breathing ecosystem teeming with potential. And it's up to us to nurture it, to replenish its nutrients, and to ensure its vitality for generations to come.

At the heart of soil health and nutrient management lies the practice of crop rotation; a time-honored tradition that harnesses the power of diversity to enrich the soil and optimize crop yields. But what exactly does this mean, and why is it so important?

Think of soil health as the foundation of a house. Just as a sturdy foundation provides stability and support for the structure above, healthy soil provides the essential nutrients and structure that plants need to thrive. When soil is rich in organic matter, teeming with beneficial microbes, and balanced in its nutrient composition, plants can grow strong and resilient, better able to withstand pests, diseases, and environmental stresses.

Crop rotation plays a crucial role in maintaining soil health by replenishing nutrients and preventing nutrient depletion. Different crops have different nutrient requirements, and by rotating between them, farmers can ensure that the soil remains balanced and fertile. For example, leguminous crops like peas and beans have the remarkable ability to fix nitrogen from the air and convert it into a form that plants can use. By planting legumes in rotation with nitrogen-hungry crops like corn or lettuce, farmers can naturally boost soil fertility without relying on synthetic fertilizers.

Crop rotation isn't just about adding nutrients to the soil; it's also about building soil structure and promoting microbial diversity. When we rotate crops, we disturb the soil in different ways, stimulating the growth of beneficial microbes and improving soil aeration and drainage. This, in turn, creates a more hospitable environment for plant roots, allowing them to access nutrients more efficiently and grow deeper and stronger.

Most importantly, crop rotation fosters a sense of connection and reciprocity with the land. When we care for the soil, it cares for us in return, providing us with nourishment and sustenance in abundance. By practicing crop rotation, we become not just farmers, but stewards of the earth; guardians of its precious resources and champions of sustainability.

So, as you embark on your journey to cultivate the land, remember the timeless wisdom of soil health and nutrient management through crop rotation. Embrace diversity, nurture the soil, and cultivate a legacy of abundance for generations to come. By nourishing the earth, we nourish ourselves and all life on this planet.

1.3 PEST AND DISEASE MANAGEMENT THROUGH CROP ROTATION

Picture a field ripe with the promise of a bountiful harvest – rows of vibrant crops reaching towards the sun, their leaves dancing in the breeze. But lurking beneath this idyllic scene lies a silent threat – pests and diseases waiting to strike, threatening to decimate our precious crops and rob us of our livelihoods. How do we protect our fields from these invisible invaders? The answer lies in the age-old practice of pest and disease management through crop rotation.

At its core, pest and disease management is about maintaining a delicate balance between the needs of our crops and the threats they face from pests and diseases. Just as we take precautions to safeguard our homes from intruders, so too must we take proactive measures to protect our crops from harm. Crop rotation offers a natural and sustainable solution, harnessing the power of diversity to disrupt pest and disease cycles and minimize their impact on our fields.

How does crop rotation help to manage pests and diseases? The answer lies in the simple yet profound principle of diversity. When we rotate between different types of crops, each with its own unique set of vulnerabilities and defenses, we create an environment that is less hospitable to pests and diseases. By breaking up monocultures; fields planted with a single crop; we reduce the risk of pest and disease buildup, making it harder for them to gain a foothold and spread throughout the field.

Take, for example, the case of the infamous corn borer, a pest that can wreak havoc on corn crops if left unchecked. By rotating corn with crops like soybeans or alfalfa, which are less susceptible to corn borers, farmers can disrupt the pest's life cycle and reduce its population in the field. Similarly, rotating crops with different root structures can help break up soil-borne disease cycles, preventing pathogens from building up in the soil and infecting susceptible crops.

Perhaps the most compelling aspect of pest and disease management through crop rotation is its sustainability. Unlike chemical pesticides and fungicides, which can harm beneficial insects and disrupt delicate ecosystems, crop rotation offers a natural and holistic approach to pest and disease control. By working with nature rather than against it, we can create healthier, more resilient fields that are better able to withstand environmental stresses and produce abundant harvests year after year.

So, as you stand in your fields, ready to defend your crops against the threats that lurk in the shadows, remember the power of crop rotation as a tool for pest and disease management. Embrace diversity, disrupt pest and disease cycles, and cultivate a legacy of resilience and abundance for generations to come. By protecting our fields today, we ensure a brighter and more sustainable future for all.

CHAPTER 2: TRADITIONAL AND MODERN CROP ROTATION SYSTEMS

2.0 TRADITIONAL CROP ROTATION PRACTICES

Close your eyes and imagine yourself stepping back in time, surrounded by the sights, sounds, and smells of a bygone era. You find yourself in the midst of a vibrant agricultural community, where farmers toil the land with weathered hands and hearts brimming with reverence for the earth. Here, amidst the ancient rhythms of the seasons, lies the timeless wisdom of traditional crop rotation practices; a legacy passed down through generations, rich with insights into the intricate dance between humanity and the land.

Traditional crop rotation practices are more than just farming techniques; they are a testament to the resilience and ingenuity of our ancestors, who worked in harmony with nature to nurture the soil and sustain life. These practices are rooted in deep observation and intimate knowledge of the land, honed over centuries of trial and error, adaptation, and innovation.

One of the most iconic examples of traditional crop rotation is the three-field system, widely practiced in medieval Europe.

In this system, fields were divided into three sections: one planted with a winter grain like wheat or rye, one with a spring grain like barley or oats, and one left fallow to rest and regenerate. This simple yet effective rotation cycle not only ensured a steady supply of food for both humans and animals but also replenished the soil with essential nutrients and prevented soil erosion.

In other parts of the world, traditional crop rotation practices took on different forms, reflecting the unique environmental conditions, cultural traditions, and agricultural techniques of diverse communities. In the Andean highlands of South America, for example, indigenous farmers practiced a form of rotational agriculture known as "qachun waqachi," in which crops like potatoes, quinoa, and maize were rotated in terraced fields to optimize water usage and soil fertility.

Regardless of their specific form, traditional crop rotation practices share a common thread: a deep reverence for the land and a commitment to stewardship that transcends generations. These practices are grounded in the principles of sustainability, resilience, and community, embodying a holistic approach to farming that honors the interconnectedness of all living things.

Today, as we confront the challenges of climate change, soil degradation, and food insecurity, the importance of traditional crop rotation practices has never been greater. By embracing the wisdom of our ancestors and learning from their experiences, we can cultivate healthier, more resilient agricultural systems that nourish both people and planet.

So, as you walk through your fields, take a moment to pause and reflect on the legacy of traditional crop rotation practices. Honor the wisdom of generations past, and let their teachings guide you as you work to build a more sustainable and harmonious future. By embracing tradition, we honor the earth and ensure a legacy of abundance for generations to come.

2.1 THREE-FIELD SYSTEM AND OTHER HISTORICAL METHODS

Step back in time with me, to an era when the rhythms of life were intimately intertwined with the cycles of the earth. Picture yourself in a medieval village, surrounded by fields of golden grain and lush green meadows. Here, amidst the hustle and bustle of agrarian life, lies the heart of traditional agriculture; a rich tapestry woven with the threads of wisdom, innovation, and community.

One of the most iconic and enduring examples of traditional crop rotation is the three-field system, which flourished throughout medieval Europe. In this system, fields were divided into three sections: one planted with a winter grain like wheat or rye, one with a spring grain like barley or oats, and one left fallow to rest and regenerate. This simple yet ingenious rotation cycle not only ensured a steady supply of food for both humans and animals but also replenished the soil with essential nutrients and prevented soil erosion.

The three-field system was just one of many historical methods of crop rotation practiced by our ancestors. In ancient Mesopotamia, the birthplace of agriculture, farmers practiced a form of crop rotation known as "mixed cropping," where different crops like barley, wheat, and legumes were planted together in the same field to optimize yields and promote soil fertility. Similarly, in ancient China, farmers practiced a system of crop rotation known as "relay cropping," where crops were planted sequentially in the same field to maximize productivity and minimize labor.

Across cultures and continents, from the fertile plains of Egypt to the terraced hillsides of the Andes, traditional farmers developed a myriad of innovative and sustainable methods of crop rotation tailored to their unique environmental conditions and agricultural practices. These methods were not just techniques for maximizing yields; they were expressions of a profound reverence for the land and a deep understanding of the interconnectedness of all living things.

Today, as we confront the challenges of climate change, soil degradation, and food insecurity, the lessons of history are more relevant than ever. By studying the traditional methods of crop rotation practiced by our ancestors, we can glean insights into how to build more resilient and sustainable agricultural systems for the future. By embracing the wisdom of the past and combining it with modern innovations, we can cultivate a legacy of abundance for generations to come.

So, as you walk through your fields, take a moment to pause and reflect on the rich tapestry of traditional agriculture. Honor the ingenuity of our ancestors, and let their timeless teachings guide you as you work to nourish the land and sustain life.

2.2 MODERN ADAPTATIONS AND INNOVATIONS IN CROP ROTATION

In the intricate dance of agriculture, crop rotation emerges as a time-honored choreography, weaving sustainability, soil health, and productivity into a seamless routine. Yet, within the realm of this traditional practice, modern adaptations and innovations have emerged as the avant-garde, reshaping the landscape of farming for a greener, more resilient future.

Rolling fields stretching as far as the eye can see, each patch a vibrant tapestry of life. This is the canvas upon which modern farmers paint their masterpiece of sustainable agriculture. In the past, crop rotation followed a rigid schedule, often dictated by tradition and the limitations of the land. But now, armed with science and technology, farmers have become choreographers of the soil, orchestrating a symphony of crops that harmonize with the natural rhythms of the earth.

One of the most compelling innovations in modern crop rotation is the concept of precision farming. Imagine a farmer equipped with satellite imagery and soil sensors, mapping out every contour of the land with surgical precision. Armed with this data, they can tailor their crop rotation plans to the unique needs of each square meter, optimizing yields while minimizing environmental impact.

Modernization doesn't stop there. In the quest for sustainability, farmers are turning to cover crops as the unsung heroes of crop rotation. These hardworking plants not only protect the soil from erosion and nutrient depletion but also act as natural fertilizers, replenishing the earth with vital nutrients for future crops. It's a win-win solution that speaks volumes about the ingenuity of modern agriculture.

Let's not forget about the role of innovation in crop genetics. Through the marvels of biotechnology, scientists are breeding crops that are not only more resilient to pests and diseases but also better suited to the rigors of modern crop rotation systems. Imagine fields of corn that require fewer inputs and produce higher yields, or wheat varieties that thrive in diverse cropping systems without sacrificing quality. These are the fruits of innovation that are reshaping the future of agriculture.

Perhaps the most compelling argument for modern adaptations in crop rotation lies in its potential to mitigate the impacts of climate change. As extreme weather events become increasingly common, farmers must adapt or perish. By diversifying their crop rotations and embracing resilient varieties, they can weather the storms of uncertainty with confidence, ensuring a bountiful harvest for generations to come.

Modern adaptations and innovations in crop rotation represent the cutting edge of sustainable agriculture.

CHAPTER 3: PLANNING AND IMPLEMENTING CROP ROTATION

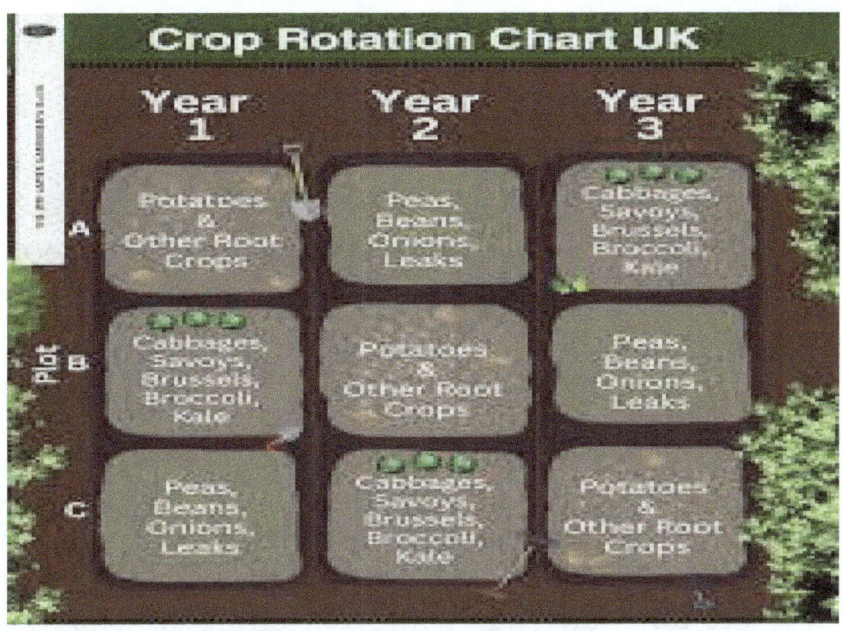

3.0 FACTORS INFLUENCING CROP ROTATION DECISIONS

The decision of which crops to rotate isn't just a matter of picking names from a hat; it's

50

a nuanced dance influenced by a myriad of factors, each playing a crucial role in shaping the farmer's strategy for success. Let's delve into the rich tapestry of considerations that sway these decisions, painting a picture of wisdom and pragmatism born from generations of agricultural knowledge.

At the heart of every crop rotation decision lies the soil itself; the very foundation upon which the farmer's livelihood depends. Soil health is paramount, and factors such as nutrient levels, pH balance, and soil structure all weigh heavily on the choice of crops. A wise farmer knows that each crop interacts with the soil in its own unique way, depleting certain nutrients while enriching others. By rotating crops strategically, they can ensure that the soil remains fertile and resilient, ready to support bountiful harvests year after year.

Soil health is just the tip of the iceberg. Climate and weather patterns also play a starring role in the crop rotation saga. From the scorching heat of summer to the icy grip of winter, each season brings its own challenges and opportunities. A savvy farmer considers these factors carefully, choosing crops that are well-suited to the local climate and capable of weathering whatever Mother Nature throws their way. After all, a crop rotation plan that ignores the whims of the weather is like sailing into a storm without a compass; a recipe for disaster.

Economic considerations also loom large in the minds of farmers as they chart their course through the seasons. Market demand, input costs, and potential profits all influence the decision-making process, guiding farmers towards crops that offer the greatest return on investment. But it's not just about maximizing profits; it's also about sustainability and long-term viability. A wise farmer knows that a diverse crop rotation can spread risk and buffer against market fluctuations, ensuring a steady income even in turbulent times.

Let's not forget about the human element. Farming isn't just a job; it's a way of life, deeply rooted in tradition and community. Family heritage, cultural practices, and local knowledge all shape the farmer's approach to crop rotation, infusing it with a sense of continuity and belonging that transcends generations. A farmer's decision to rotate crops isn't just about what makes sense on paper; it's about honoring the wisdom of those who came before and paving the way for those who will follow.

The factors influencing crop rotation decisions are as diverse and multifaceted as the crops themselves. From soil health to climate resilience, economic viability to cultural heritage, each consideration plays a vital role in shaping the farmer's strategy for success. It's a delicate balancing act, guided by wisdom, pragmatism, and a deep respect for the land. And in the end, it's this timeless blend of tradition and innovation that ensures a bountiful harvest for generations to come.

3.1 CROP SELECTION CRITERIA FOR ROTATION SYSTEMS

Choosing the right crops for rotation isn't just about picking names from a list; it's a thoughtful and strategic process that can make or break a harvest. Let's dive deep into the criteria that guide farmers as they navigate this crucial decision-making journey, weaving together wisdom, experience, and a deep respect for the land.

First and foremost, soil health takes center stage in the crop selection process. Just as a house needs a strong foundation, crops rely on healthy soil to thrive. Farmers look at factors like nutrient levels, pH balance, and soil structure to determine which crops will best complement and nourish the earth. By rotating crops strategically, they can replenish nutrients, suppress pests and diseases, and improve overall soil fertility, ensuring a vibrant and resilient ecosystem for future generations.

Soil health is just one piece of the puzzle. Climate and weather patterns also play a pivotal role in crop selection. From blistering heatwaves to bone-chilling frosts, each season brings its own set of challenges and opportunities. Farmers consider factors like temperature, rainfall, and growing season length when deciding which crops to rotate, choosing varieties that are well-suited to the local climate and capable of thriving in diverse conditions. After all, a successful crop rotation plan is like a well-tailored wardrobe; it adapts to the seasons and always looks good on the farm.

Economic considerations also weigh heavily on the minds of farmers as they ponder their crop selection. Market demand, input costs, and potential profits all factor into the decision-making process, guiding farmers towards crops that offer the greatest return on investment. But it's not just about making money; it's also about sustainability and long-term viability. A diverse crop rotation can spread risk and buffer against market fluctuations, ensuring a stable income even in uncertain times.

Let's not forget about the human element. Farming isn't just a job; it's a way of life, deeply rooted in tradition and community.

Family heritage, cultural practices, and local knowledge all shape the farmer's approach to crop selection, infusing it with a sense of continuity and belonging that transcends generations. A farmer's decision to rotate crops isn't just about what makes sense on paper; it's about honoring the wisdom of those who came before and paving the way for those who will follow.

Crop selection criteria for rotation systems are as diverse and multifaceted as the crops themselves. From soil health to climate resilience, economic viability to cultural heritage, each consideration plays a vital role in shaping the farmer's strategy for success. It's a delicate balancing act, guided by wisdom, experience, and a deep love for the land. And in the end, it's this timeless blend of tradition and innovation that ensures a bountiful harvest for generations to come.

3.2 DESIGNING CROP ROTATION SEQUENCES FOR MAXIMUM BENEFIT

In the art of agriculture, there exists a fundamental principle that has stood the test of time: crop rotation. Dating back to ancient civilizations, this practice involves alternating the types of crops grown in a particular field season after season. Yet, in today's modern farming landscape, the importance of designing crop rotation sequences for maximum benefit cannot be overstated. It's not just about tradition; it's about sustainable agriculture, soil health, pest management, and ultimately, maximizing yields.

The Benefits of Crop Rotation:
1. Soil Health: Different crops have different root structures and nutrient requirements. By rotating crops, farmers can prevent soil

depletion and promote soil health by replenishing nutrients and organic matter.

2. Pest and Disease Management: Continuous monoculture cropping can lead to the buildup of pests and diseases that target specific crops. By rotating crops, farmers can disrupt these cycles, reducing the need for chemical pesticides and herbicides.

3. Weed Control: Certain crops, like legumes, can suppress weeds through shading and allelopathy. By incorporating these crops into rotation sequences, farmers can naturally manage weed populations.

4. Improved Yields: A well-designed crop rotation sequence can lead to increased yields over time by optimizing nutrient utilization, reducing pest pressure, and improving soil structure.

Designing Crop Rotation Sequences:

1. Crop Selection: Choose crops that complement each other in terms of nutrient needs, growth habits, and pest and disease susceptibility. Consider factors such as climate, soil type, and market demand when selecting crops.

2. Rotation Period: Determine the length of the rotation cycle based on the specific needs of the crops chosen and the desired outcomes. A typical rotation cycle may range from two to five years.

3. Cover Crops: Incorporate cover crops into rotation sequences to improve soil structure, suppress weeds, and provide additional nutrients. Legumes, such as clover and peas, are excellent choices for cover crops due to their ability to fix nitrogen.

4. Green Manure: Utilize green manure crops, such as rye or vetch, to add organic matter to the soil and improve fertility. Green manure crops can be incorporated into rotation sequences or used as winter cover crops.

5. *Crop Residues:* Leave crop residues on the soil surface to protect against erosion, conserve moisture, and provide organic matter for soil microorganisms.

Designing crop rotation sequences for maximum benefit is akin to composing a symphony of sustainability. It requires foresight, planning, and a deep understanding of the land. By embracing the principles of crop rotation, farmers can cultivate not only bountiful harvests but also a healthier, more resilient ecosystem for generations to come.

CHAPTER 4: CROP ROTATION STRATEGIES FOR DIFFERENT AGROECOSYSTEMS

4.0 CROP ROTATION IN INTENSIVE AGRICULTURE

In the bustling world of intensive agriculture, where the demands for higher yields and sustainability echo louder than ever, one age-old practice stands out as a beacon of hope: crop rotation. Far from being a mere tradition, crop rotation is a dynamic strategy that breathes life into our soils, empowers farmers, and safeguards our food security for generations to come.

Soil Health: The Foundation of Agricultural Success

Picture a flourishing farm where the soil teems with life ; earthworms tunneling through its depths, microorganisms bustling with activity, and roots delving deep to access vital nutrients. This is the magic of crop rotation. By alternating crops with different nutrient needs, farmers can prevent soil depletion, maintain its structure, and enhance its fertility naturally.

Pest and Disease Management

Intensive agriculture often faces a formidable foe: pests and diseases. But rather than waging war with chemicals, crop rotation offers a sustainable solution. By disrupting the lifecycle of pests and pathogens, it reduces their buildup in the soil, thus curbing outbreaks and minimizing the need for pesticides. It's a win-win for farmers and the environment alike.

Yield Maximization

In the fast-paced world of intensive agriculture, every harvest counts. Crop rotation is the secret weapon that ensures consistent yields year after year. By harnessing the complementary relationships between crops, it optimizes resource use, reduces input costs, and boosts overall productivity. It's a testament to the power of working with nature, not against it.

Adaptability to Climate Change

As the specter of climate change looms large, farmers face increasingly unpredictable growing conditions. Crop rotation emerges as a resilient strategy, offering flexibility in the face of adversity. By diversifying crop types, farmers can hedge their bets against extreme weather events, ensuring a more stable and secure food supply for all.

Empowering Farmers

Beyond its agronomic benefits, crop rotation empowers farmers to take control of their livelihoods. By fostering diversity in their fields, they reduce their reliance on external inputs, enhance their resilience to market fluctuations, and cultivate a deeper connection to the land they steward. It's a model of agriculture that honors both tradition and innovation.

In the dynamic landscape of intensive agriculture, crop rotation shines as a beacon of sustainability

4.1 CROP ROTATION IN ORGANIC FARMING SYSTEMS

Step into the world of organic farming, where the earth sings with life, and every seed planted is a promise of abundance. At the heart of this sustainable agriculture lies a timeless practice: crop rotation. More than just a technique, it's a dance with nature, a symphony of soil, sun, and seasons. In the gentle rhythm of crop rotation, organic farmers find not only sustenance but a profound connection to the land and the communities they nourish.

Nurturing the Soil

In organic farming, the soil is not just dirt but a living, breathing ecosystem teeming with microscopic life. Crop rotation is the gentle touch that rejuvenates this vital foundation. By alternating crops with different nutrient needs and root structures, organic farmers replenish the soil, improving its fertility and structure naturally. It's a cycle of giving and receiving, where the land flourishes, and so do we.

Balancing Pests and Predators

In the delicate balance of organic ecosystems, pests and predators play their roles, but harmony is key. Crop rotation offers a natural solution, disrupting pest life cycles and creating habitats for beneficial insects. Ladybugs, lacewings, and spiders become allies in the fight against pests, reducing the need for synthetic pesticides and preserving the health of our environment and communities.

Diversifying Yields

In the tapestry of organic farming, diversity is the thread that binds us to the land and each other. Crop rotation celebrates this diversity, weaving together a mosaic of crops with varied needs and attributes. By rotating legumes, grains, vegetables, and cover crops, organic farmers optimize yields, minimize risks, and create a symphony of flavors and nutrients that nourish both body and soul.

Climate Resilience

As the specter of climate change looms large, organic farmers stand as stewards of resilience and adaptation. Crop rotation is their secret weapon, a shield against the storms and droughts that threaten our food supply. By nurturing healthy soils and diverse ecosystems, organic farmers create buffers against extreme weather events, ensuring a more stable and secure future for all.

Cultivating Connection

Beyond its practical benefits, crop rotation fosters a deeper connection between farmers, consumers, and the land. In the shared rhythm of planting and harvest, we find common ground and a sense of belonging to something greater than ourselves. Organic farming becomes not just a livelihood but a way of life, rooted in tradition, nourished by community, and sustained by love for the land.

In the vibrant tapestry of organic farming, crop rotation emerges as a cornerstone of sustainability, resilience, and connection. It's a testament to the power of working with nature, not against it, and a reminder that the earth is not just a resource to be exploited but a partner to be cherished and respected. As we strive to build a more sustainable and equitable food system, let us embrace the art of crop rotation and sow the seeds of a brighter future for all.

4.2 CROP ROTATION IN PERENNIAL CROPPING SYSTEMS

Step into the world of perennial cropping systems, where trees sway in the breeze, vines climb toward the sun, and the land is alive with the promise of perennial harvests. Amidst this bounty, one practice stands out as a guiding principle: crop rotation. Far from being a fleeting trend, it's a timeless strategy that honors the rhythms of nature, fosters soil health, and sustains livelihoods for generations to come.

Embracing Diversity:

In perennial cropping systems, diversity is the cornerstone of resilience. Crop rotation takes on a new dimension as farmers interplant trees, shrubs, vines, and perennial herbs in intricate patterns. By rotating crops with different root depths, nutrient requirements, and growth habits, they create a vibrant tapestry of life that nurtures the soil, supports beneficial organisms, and maximizes yields year after year.

Nurturing Soil Health:

The soil is the silent hero of perennial cropping systems, anchoring roots, storing water, and cycling nutrients through the ecosystem. Crop rotation becomes a dance of renewal, as each crop contributes its unique blend of organic matter, microbes, and minerals to the soil. By alternating between nitrogen-fixing legumes, deep-rooted perennials, and nutrient-rich cover crops, farmers build soil fertility and structure, laying the foundation for bountiful harvests and healthy ecosystems.

Managing Pests and Diseases:

In the lush landscape of perennial crops, pests and diseases can pose a challenge, but crop rotation offers a natural defense. By diversifying plantings and interrupting pest life cycles, farmers create habitats for beneficial insects and microbes that help keep pests in check. It's a delicate balance, where predators and prey coexist in harmony, and chemical pesticides become a last resort rather than a first line of defense.

Sustainable Resource Management:

Perennial cropping systems are not just about growing food; they're about stewarding resources for future generations. Crop rotation plays a vital role in this sustainability, helping farmers manage water, nutrients, and energy more efficiently. By integrating trees and shrubs into crop rotations, farmers enhance water retention, reduce erosion, and sequester carbon, mitigating the impacts of climate change and preserving the health of our planet.

Cultivating Community:
Beyond its agronomic benefits, crop rotation fosters a sense of connection and belonging in perennial cropping systems. In the shared rhythm of planting and harvest, farmers come together to share knowledge, resources, and stories passed down through generations. Perennial cropping becomes not just a livelihood but a way of life, rooted in tradition, nourished by community, and sustained by a deep reverence for the land.

In perennial cropping systems, crop rotation emerges as a beacon of resilience, sustainability, and community. It's a testament to the power of working with nature, not against it, and a reminder that the earth is not just a resource to be exploited but a partner to be cherished and respected. As we strive to build a more sustainable and equitable food system, let us embrace the art of crop rotation and sow the seeds of a brighter future for all.

CHAPTER 5: INTEGRATING LIVESTOCK IN CROP ROTATION

5.0 ROLE OF LIVESTOCK IN SUSTAINABLE CROP ROTATION

Rolling fields dotted with vibrant crops swaying gently in the breeze, while contented livestock graze nearby. This idyllic scene isn't just picturesque; it's the epitome of sustainable agriculture in action. Livestock, often overshadowed by the allure of crops, are the unsung heroes of this agricultural ballet. Let's uncover the vital role they play in the intricate tapestry of sustainable crop rotation.

Enhancing Soil Health:

At the heart of any successful farming endeavor lies the soil. It's not just dirt; it's the lifeblood of agriculture. Livestock, with their natural grazing behaviors, contribute significantly to soil health. As they munch on grasses and forage, they inadvertently aerate the soil with their hooves, promoting better water infiltration and root growth. Moreover, their manure acts as a natural fertilizer, enriching the soil with essential nutrients and organic matter. It's a symbiotic relationship where both crops and livestock thrive in harmony.

Pest and Weed Management:

Anyone who's battled with weeds knows the struggle is real. But imagine having natural allies in the fight against these pesky intruders. Livestock, with their insatiable appetites, serve as biological weed controllers. By grazing on unwanted plants, they help keep weed populations in check, reducing the need for herbicides and manual labor. Additionally, their presence disrupts the life cycles of pests, creating a natural barrier that safeguards crops without harming the environment.

Diversifying Income Streams:

In today's unpredictable agricultural landscape, diversification is key to resilience. Livestock provide farmers with additional income streams beyond traditional crop sales. Whether it's selling meat, milk, wool, or other animal products, livestock offer a buffer against market fluctuations and economic uncertainties. This diversification not only spreads risk but also adds depth to farming operations, ensuring financial stability for generations to come.

Fostering Community Connections:

Farming isn't just a profession; it's a way of life deeply rooted in community and tradition. Livestock, with their charming personalities and enduring presence, serve as catalysts for connection. From livestock shows and fairs to community-supported agriculture programs, they bring people together, fostering a sense of camaraderie and belonging. Moreover, the shared responsibility of caring for animals strengthens bonds within families and communities, creating a support network that extends far beyond the farm gates.

Preserving Biodiversity:

In an era of monoculture and genetic homogenization, preserving biodiversity is more critical than ever. Livestock, particularly heritage breeds, play a crucial role in safeguarding genetic diversity within agricultural systems. By raising and breeding diverse livestock breeds, farmers contribute to the preservation of rare and endangered species, ensuring a resilient gene pool for future generations. Additionally, integrating livestock into crop rotation fosters a more balanced ecosystem, where plants, animals, and microorganisms coexist in harmony.

As stewards of the land, it's our duty to recognize and celebrate the indispensable role of livestock in sustainable crop rotation. After all, a thriving farm is not just measured by its yields but by the harmony it creates within the natural world and the communities it serves.

5.1 GRAZING AND FORAGE MANAGEMENT IN CROP ROTATION SYSTEMS

Step onto any farm, and you'll witness a delicate symphony of life unfolding before your eyes. Grazing animals and lush forage play the leading roles, weaving together the melody of sustainable agriculture. Let's explore the captivating world of grazing and forage management in crop rotation systems, where every step on the land is a testament to our stewardship of nature's bounty.

Creating Healthy Soils:
Beneath the surface of the earth lies a hidden world teeming with life, where soil health reigns supreme. Grazing animals, with their gentle tread and voracious appetites, are nature's architects of soil fertility. As they graze on pasturelands, they stimulate plant

growth and trample organic matter into the soil, enriching it with vital nutrients. Their manure, a precious gift from nature, nourishes the earth and fosters the growth of healthy, resilient crops in subsequent rotations. It's a symbiotic relationship where the land gives back what it receives, ensuring a legacy of abundance for generations to come.

Balancing Ecosystem Dynamics:
In the whole form of nature, balance is key. Grazing animals, when managed thoughtfully, serve as stewards of biodiversity and habitat restoration. By grazing on a diverse array of plants, they prevent the dominance of invasive species and promote the growth of native vegetation. This diversity not only supports a multitude of wildlife but also creates resilient ecosystems that can withstand the impacts of climate change. It's a testament to the power

of nature's wisdom, where every species has a role to play in the symphony of life.

Mitigating Weed and Pest Pressure:
Weeds and pests are the perennial foes of farmers everywhere. Yet, instead of resorting to chemical warfare, why not enlist the help of nature's allies? Grazing animals, with their insatiable appetites, excel at weed and pest management. By grazing on undesirable plants and disrupting pest life cycles, they reduce the need for harmful pesticides and herbicides. It's a natural solution that not only protects crops but also preserves the delicate balance of the ecosystem.

Enhancing Climate Resilience:
In an era of climate uncertainty, resilience is paramount. Grazing and forage management offer a beacon of hope in the face of climate change. Well-managed pasturelands act as carbon sinks, sequestering atmospheric

carbon and mitigating the impacts of greenhouse gas emissions. Additionally, diverse pasture ecosystems are more resilient to extreme weather events, providing a buffer against droughts, floods, and other climate-related challenges. It's a testament to the adaptability of nature, where diversity fosters strength and resilience.

Fostering Economic Prosperity:
At the heart of every farm lies the desire for economic prosperity. Grazing and forage management offer a pathway to sustainable profitability. By reducing input costs on fertilizers and pesticides and providing additional income streams through livestock production, they ensure financial stability for farmers and rural communities. Moreover, diversified farming operations are more resilient to market fluctuations, safeguarding livelihoods and ensuring a vibrant

agricultural landscape for generations to come.

Grazing and forage management are the orchestrators of harmony and balance. It's our collective responsibility to embrace the art of grazing and forage management in crop rotation systems and cultivate a future where nature's symphony resounds with abundance and vitality.

5.2 MANURE MANAGEMENT AND NUTRIENT CYCLING

In the very aspect of agriculture, the relationship between livestock and crops is often overlooked. Yet, when we delve deeper into the ecosystem of farming, we uncover the vital role of manure management and nutrient cycling. This symbiotic relationship holds the key to sustainable farming practices, ensuring the health of both the land

and its inhabitants. Let us embark on a journey to understand the significance of integrating livestock in crop rotation and how thoughtful manure management can revolutionize our approach to agriculture.

Understanding Manure Management:

Manure, often dismissed as waste, is a treasure trove of nutrients essential for soil fertility. Rich in nitrogen, phosphorus, potassium, and organic matter, properly managed manure can rejuvenate tired soils, enhance crop yields, and reduce reliance on synthetic fertilizers. However, effective manure management goes beyond mere application; it requires careful planning, proper storage, and strategic utilization.

Nutrient Cycling in Agriculture:

At the heart of sustainable agriculture lies the concept of nutrient cycling. Livestock graze on pastures, consuming nutrients from the soil. Through their waste, these nutrients are returned to the land, completing a natural cycle. Integrating livestock into crop rotation enhances this process, allowing for the efficient recycling of nutrients. As crops utilize these nutrients, they, in turn, provide feed for livestock, closing the loop of sustainability.

Benefits of Integrated Livestock Management:
The benefits of integrating livestock in crop rotation extend far beyond nutrient cycling. Livestock contribute to soil health by enhancing microbial activity and soil structure through their grazing patterns. Additionally, rotational grazing practices prevent soil erosion, promote biodiversity, and mitigate greenhouse gas emissions. Moreover, diversified farming systems that incorporate both crops and livestock are more

resilient to environmental fluctuations and market volatility.

Challenges and Solutions:

While the concept of integrated livestock management is appealing, it is not without challenges. Manure management can pose environmental risks if not executed properly, leading to nutrient runoff and water contamination. However, by adopting best management practices such as composting, anaerobic digestion, and precision application techniques, these challenges can be mitigated. Furthermore, investment in research and education is crucial to empower farmers with the knowledge and tools necessary for sustainable manure management.

Manure management and nutrient cycling are integral components of sustainable agriculture. Let us cultivate a future where manure is no longer viewed as waste, but rather as a valuable resource for nourishing the land and feeding the world. Together, we can pave the way towards a more resilient, regenerative, and prosperous agricultural system.

CHAPTER 6: CASE STUDIES AND SUCCESS STORIES

6.0 CASE STUDIES OF EFFECTIVE CROP ROTATION PRACTICES

The agriculture sector has been a significant part of the global industry for centuries. Farmers have continued to develop new practices in cultivating crops to improve yields, economic returns, and soil health. One such practice that has been gaining more attention lately is crop rotation. The best thing about crop rotation is that it not only enhances crop productivity, but it also nourishes soil health to improve future crop yields. Crop rotation is an excellent example of a sustainable and eco-friendly agriculture practice that has achieved remarkable results in improving yields, reducing inputs and enhancing soil health.

Effective crop rotation practices have been increasingly popular due to their benefit to the environment and food production. In a crop rotation system, farmers grow different crops in the same field in a cyclical manner, as opposed to planting the same crop repeatedly. This allows each crop to improve soil health, pest prevention, increased nutrient absorption, and water conservation. It is a much more sustainable and resilient practice that seeks to break the cycle of pest infestations and soil depletion. Crop rotation is, therefore, the keystone of good soil stewardship.

One successful example of crop rotation practice is the three-season rotation in Iowa, USA. The system consists of planting corn, soybeans, and oats in sequence. This rotation enables farmers to achieve higher yields than when each crop is planted alone. The corn plant's high nitrogen usage stimulates soybean growth, which in turn provides nitrogen to the oat plant. The oat plant then improves the soil's structure, leading to better soil moisture and fertility for future crops in the cycle.

In Ukraine, an organic farmer, Oleksander Sukhorukov, adopts the eight-shift rotation system as part of his organic farming practices. The system consists of six years of crops and two years of leaving the soil fallow in a particular field. Vegetables are the main crops, but the rotation also includes cereals, cover crops, and leys. Sukhorukov's system is not only beneficial to the environment but also economically efficient, as yields have doubled, and profits have increased significantly.

In Tanzania, a collaboration between the International Institute of Tropical Agriculture and local farmers saw the introduction of banana-peanut intercropping to farmers. This practice not only provided food and income for local farmers, but it also prevented soil degradation and erosion, leading to soil regeneration in previously barren fields. The intercropping system captures rainwater, which increases soil moisture, resulting in more yield and sustainability.

Crop rotation is a beneficial and sustainable practice that is more critical now than ever before to improve soil health and crop productivity. With climate change set to impact crops and farm produce adversely, we need to adopt more eco-friendly farming practices like crop rotation. Farmers worldwide can learn from case studies of successful crop rotation practices, such as those in Iowa, Ukraine, and Tanzania, and adapt them to their farming systems. Adopting crop rotation practices helps to save the planet, but it also makes financial sense for farmers, improves soil health, and ultimately, benefits us all.

6.1 SUCCESS STORIES FROM FARMERS IMPLEMENTING CROP ROTATION STRATEGIES

In recent years, more and more farmers have been implementing crop rotation strategies, and the results have been incredible. By regularly alternating their crops, farmers are seeing increased yields, reduced soil erosion, and improvements in soil health.

Don't just take our word for it; let's take a look at some real-world success stories from farmers who have implemented crop rotation strategies:

Meet Sarah, a farmer in the midwest of the United States. For years, Sarah had been struggling to grow her crops due to the depletion of nutrients in her soil. She decided to implement a crop rotation strategy, alternating between corn and soybeans each year. The results were incredible - not only did Sarah see a 15% increase in her yields, but she also noticed that her soil health improved drastically.

Consider the case of Li, a farmer in China. Li had been growing crops on the same plot of land for years, and he noticed that his yields were decreasing each year. He decided to try a crop rotation strategy, and began alternating between rice and a legume crop. The results were transformative - Li saw an increase in yield of up to 20%, and the nutrients in his soil were replenished.

These are just two examples of the many farmers who have seen success by implementing crop rotation strategies. And the benefits don't stop at just improved yields; crop rotation can also reduce soil erosion, decrease pest and disease pressure, and even reduce the need for fertilizers and other inputs.

Why aren't more farmers implementing crop rotation strategies? It can be a difficult process to navigate, especially for those who have been farming the same plot of land for years. But as Sarah and Li have shown us, the benefits of crop rotation are undeniable. With careful planning and a willingness to try something new, farmers can see drastic improvements in both their crops and their soil health.

The success stories of farmers implementing crop rotation strategies are truly inspiring. By taking a chance on something new, these farmers have been able to see incredible improvements in their yields, soil health, and more. It is our hope that these stories will encourage others to try crop rotation for themselves, and experience success just like Sarah and Li.

6.2 CHALLENGES AND LESSONS LEARNED

Life is never a bed of roses, and almost every individual in the world has had to face challenges at different stages of life. It is important to note that challenges are not only inevitable but also necessary for personal growth, development, and success. A person who has never encountered challenges is likely to lack the skills and experience necessary to cope with the ups and downs of life. In this article, we will explore some challenges faced by individuals in the real world and the lessons they have learned along the way.

Financial Challenges

One of the most significant challenges people face in the world today is financial instability. Many people struggle to make ends meet and maintain financial stability in the long run. Individuals who face financial challenges often learn the importance of managing their finances effectively. Through experience, they learn to budget, save, invest, and avoid taking on unnecessary debt. They also learn the importance of patience and perseverance in working towards financial stability. It is essential to note that financial challenges can also teach individuals the value of hard work and determination, as individuals work tirelessly to solve their financial problems.

Relationship Challenges

Relationship challenges are also prevalent in the world today. Whether in family settings or romantic relationships, individuals face relationship challenges that can sometimes be overwhelming. These challenges can take the form of communication breakdown, infidelity, betrayal, and a host of other issues. Individuals who overcome these challenges often learn the importance of effective communication, empathy, forgiveness, and trust. They learn that relationships require effort and commitment, and that building and maintaining healthy relationships takes time and dedication.

Career Challenges

Career challenges are another set of challenges individuals face in the real world. Individuals may face challenges such as job loss, workplace stress, or challenges associated with moving up the career ladder. Individuals who overcome these challenges often learn the importance of resilience, determination, and flexibility. They learn to adapt to change and continually develop new skills to remain relevant in their careers.

Health Challenges

Another set of challenges individuals face are health challenges. Individuals experience health challenges such as chronic illnesses, accidents, and other health issues. Individuals who overcome these challenges often learn the importance of taking care of their health, developing resilience, and finding support from family and friends. They also learn to appreciate life more, realizing that it is a precious gift that should not be taken for granted.

Challenges are an inevitable aspect of life, and no one is immune to them. However, overcoming challenges provides individuals with invaluable lessons and skills that can help them navigate future obstacles. By learning from the challenges they face, individuals develop resilience, determination, and the necessary skills to achieve personal and professional success.

CONCLUSION

THE FUTURE OF CROP ROTATION: INNOVATIONS AND OPPORTUNITIES

As we look toward the future of agriculture, one thing is certain: the practice of crop rotation will be more crucial than ever before. With a rapidly growing population and increasing pressure on our planet's limited resources, it is becoming apparent that we must innovate and optimize the methods we use to grow our food. Fortunately, crop rotation is a practice that has proven effective for centuries and continues to show promise as we adapt to meet these challenges.

For those unfamiliar with the practice, crop rotation is a technique where different crops are planted in the same field in successive growing seasons. Each crop has its unique nutrient requirements and affects the soil health in different ways. By rotating crops with different needs, farmers can mitigate soil-borne diseases and pests, improve soil fertility, prevent erosion and reduce the need for synthetic fertilizers and pesticides.

As we move forward, there are several exciting innovations and opportunities that can expand and refine the practice of crop rotation further.

Firstly, the use of cover crops has become increasingly popular in recent years. Planting cover crops like clover, rye, or barley in between growing seasons helps to reduce the loss of nutrients through soil erosion, improves soil moisture, and it also adds organic matter to the soil. Cover crops help prevent soil degradation and maintain the fertility and productivity of farming land.

Secondly, precision agriculture technology can augment our ability to implement crop rotation on a larger scale and more effectively. For instance, the use of satellite imagery and soil sensors allow farmers to address specific aspects of soil health that are elusive, like nutrient deficiencies, pH levels, and moisture content. These technologies can help farmers assess the soil's health and identify the crops that will produce the best results for their specific field.

Additionally, as sustainability and eco-friendliness concerns grow, crop rotations that include perennials, such as fruit and nut trees, grape vines, or berry bushes, can improve soil health, prevent erosion and extend the period of photosynthetic activity. Moreover, the integration of livestock into cropping systems, where crops and livestock are raised together, has also shown exciting potential. This approach brings the benefits of animal manure into the crop rotation system, providing natural fertilization while also improving soil structure.

Innovation and opportunities abound for the future of crop rotation. By using cover crops, adopting precision agriculture technologies, and integrating perennials and livestock, we can develop more sustainable, efficient, and productive methods of farming. These new approaches allow us to make the most of our land, protect our precious soil resources, and feed our growing population while conserving our planet and maintaining a healthy ecosystem. The future of crop rotation is bright, and as we continue to innovate and learn, it will only get better.

APPENDICES

CROP ROTATION PLANNING WORKSHEETS AND TOOLS

Crop rotation planning worksheets and tools are essential for any farmer or gardener who wants to maximize their yields and practice sustainable agriculture. These worksheets and tools help farmers plan out their crop rotation strategy, ensuring they have a balance of nutrients in their soil, pest management, and crop diversity.

To get started, farmers need to assess their land, soil type, and climate. This information helps to determine the best crops to plant and how often the crops should rotate. The first step is to create a crop rotation plan, which outlines the crops to be planted in a particular area and the length of time they will be grown.

Farmers can use various tools such as calendars, spreadsheets, or online tools to plan their rotation strategy. These tools can help farmers calculate the interval between crops, determine which crops to plant in each field section, and ensure that the soil is not depleted of essential nutrients.

To ensure success, it is crucial to follow an appropriate crop rotation scheme that restores soil fertility, controls pests and disease, and maximizes the land's potential. Farmers can select from various rotation options such as three-year rotations, four-year rotations, or more extended crop rotations, depending on their land's condition and their agricultural goals.

One example of a crop rotation plan is the three-year rotation. In this scheme, farmers plant legumes that fix nitrogen for the first year, followed by cereal for the second year, and then finally, they plant crucifers in the third year. This sequence replenishes the soil with nutrients, breaks the pest cycles, and increases soil biodiversity.

Effective crop rotation planning worksheets and tools are crucial for a successful farming or gardening experience. By understanding the land, climate, and soil fertility status, and using the right tools, farmers can create an appropriate crop rotation plan that maximizes crop yield, soil biodiversity, and enhances sustainability. In this way, farmers can improve the soil quality over time and produce nutrient-rich crops, which is essential for human life and the environment.

www.ingramcontent.com/pod-product-compliance
Lightning Source LLC
Chambersburg PA
CBHW070107230526
45472CB00004B/1157